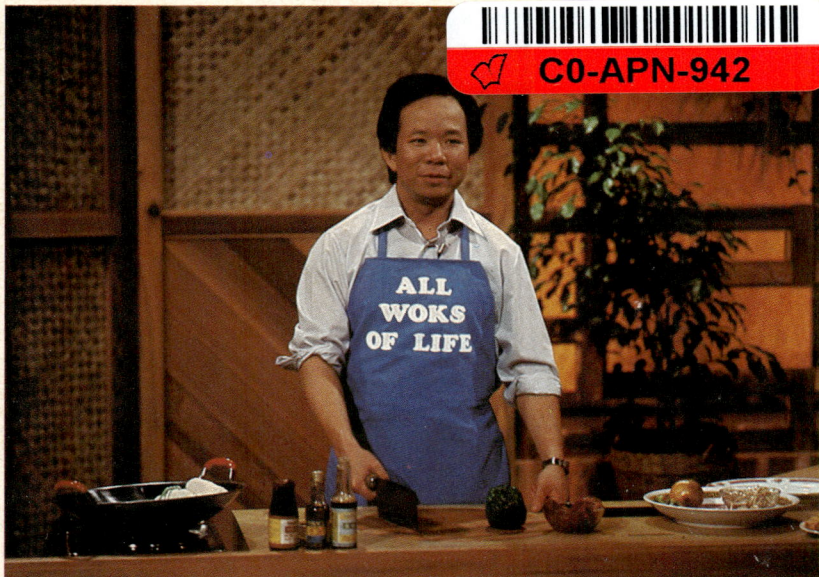

**All Woks of Life**

**Almond Jello with Fruit Cocktails**

C0-APN-942

**Wok the Heck**

**Bean Sprouts with Mixed Vegetables**

**Wok a Day Keeps Hunger Away**

**Brussels Sprouts, Stir-Fried**

**In Action the Happy Cooker**

**Mushrooms with Baby Corns**

**Wok You See is Wok You Get**

**Mushrooms in Oyster Sauce**

**Fried Rice**

**Zucchini in Black Bean Sauce**

**Tofu with Oyster Sauce**

**Stephen's Restaurant at 9948 Lougheed Hwy.
Burnaby, B.C. Canada — Tel: (604) 421-4901**

**Inside Yan's Chinese Restaurant**

# VEGETABLES

## THE CHINESE WAY

# 菜 譜

**by**

### STEPHEN YAN

| | |
|---|---|
| 1st Edition | January 1978 |
| 2nd Edition | August 1978 |
| 3rd Edition | March 1979 |
| 4th Edition | March 1980 |
| 5th Edition | November 1980 |
| 6th Edition | July 1981 |

### PUBLISHED BY

## YAN'S VARIETY COMPANY LIMITED

### P.O. BOX 227

### PORT COQUITLAM, B.C. V3C 3V7

### CANADA

**All rights reserved; no part of this book may be reproduced or duplicated in any form without the written permission of the publisher.**

© **Copyright 1981 by YAN'S**
**Printed in Canada**

Let's Wok Together

# ABOUT THE AUTHOR

Born and raised in Hong Kong, Mr. Stephen Yan received his training and practice in authentic Chinese cooking since the age of ten. He came to North America in 1963 and has taught thousands of students through his cooking classes, department store demonstrations, radio and television shows.

Mr. Stephen Yan, as the Director of Yan's conglomerate is:
—the President of Yan's Variety Company Ltd., which manufacturers Chinese cookwares and condiments supplying hundreds of outlets throughout Canada;
—the Principal of Yan's Gourmet Chinese Cooking School in Burnaby, B.C., Canada;
—the Chief Chef at Yan's Gourmet Chinese Restaurant located in Burnaby, B.C., Canada;
—the hosting star and producer of the popular Chinese cooking television shows, "YAN'S WOKING" on British Columbia Television Broadcasting System Ltd., and "WOK WITH YAN" on the Canadian Broadcasting Corporation National Network aired across Canada and northern parts of the United States.

Mr. Yan's active participation in public appearances, including the Alan Hamel Show and his extensive travel throughout China, Hong Kong, South East Asia and North America, has made him a popular celebrity. Mr. Yan has also hosted several Chinese New Year Festivals, Home Shows, and Gourmet Shows in Canada. He is a feature food writer for several newspapers in Canada and is the author of FIVE "BEST SELLER" Chinese Cookbooks.

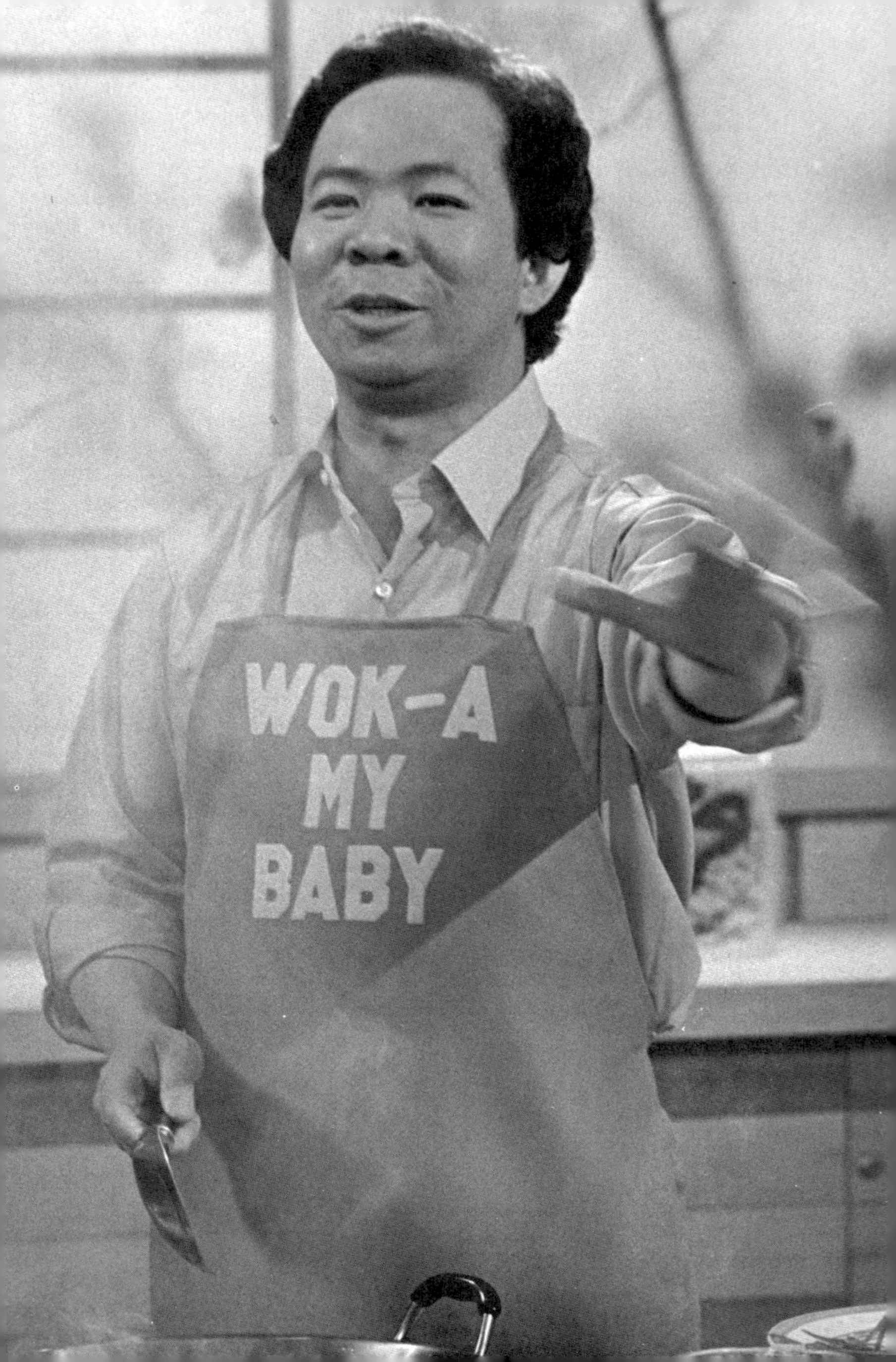
WOK-A
MY
BABY

# FOREWORD

"A country exists with people, and people live on food." This old Chinese saying emphasizes the importance of good meal in our daily life. The desire of cooking, the choice of food and the way a meal is consumed can represent the physical and spiritual well-being of a person. An unhealthy stomach can not tolerate spicy food, and a person with an unsettled mind bothers with no meal.

Cooking is healthy, recreational, fun and should be enjoyed. It keeps our mind occupied; it motivates interest and imagination; it brings family and friends together; it gives us a great sense of accomplishment and self-satisfaction; and most of all, a successfully prepared meal can certainly promote a better health for our body and mind.

This book is designed for those who have a desire and courage to learn cooking. There is no mystery in Chinese Cooking. We all learn from practising and experimenting in the kitchen. Every dish is prepared with imagination and trials. In this book, recipes are straight forward, practical, extremely "FLEXIBLE" to suit your taste and materials at hand. It is written in such a way that if certain ingredients are not available substitution can be used without seriously changing the flavour and result. Therefore, changes in recipes to meet your needs are encouraged as you may prefer more of one item and less of another. By following the technique and steps, good results can be expected. Through years of teaching experience, the recipes in this book have been chosen and tested for the consideration of:

      (A)    Step by step learning progress
      (B)    Ease of performance
      (C)    Compatibility in taste
      (D)    Use of available ingredients

For those who make use of this book, I wish every happiness and success in cooking; good health in your body and mind.

    HAPPY COOKING!!

                      Stephen Yan

Vancouver,
British Columbia, CANADA

# INTRODUCTION

Eating is one of the basic activities in our daily lives. Eating well has always been the aspiration of most societies. Modern living has emphasized this especially through the media of television and especially in the more affluent societies in the Western Hemisphere. With this has come an increasing incidence of nutritional diseases, namely generalized increase in obesity, a high incidence of heart disease especially coronary artery disease, diabetes mellitus and hypertension as the more notable diseases.

Now we have come to the realization that not only do we have to eat well but also eat wisely and nutritiously and more experts both culinary and medical are emphasizing low calorie, low cholesterol and low carbohydrate diets.

Chinese cooking has traditionally and basically always centered around these very concepts of low calorie, low cholesterol, making full use in exploiting the natural flavors of vegetables and other low cholesterol foods using very effective cooking methods that are presently being used to great advantage by the new wave of contemporary French chefs who are responsible for the new French cuisine of the 70's.

Stephen Yan has established a justified reputation as an outstanding teacher and guide to the art of Chinese cooking with his very popular Chinese cooking classes in the lower Mainland of British Columbia and numerous television appearances, newspaper articles across Canada and the United States. Thousands of enthused and satisfied students and their families are enjoying the gourmet delights of Chinese cuisine, and also making full use of the flexible method being taught by Mr. Yan giving full scope for improvisation.

Now Mr. Yan has taken the art of Chinese cooking to the next logical step by his new book "VEGETABLES THE CHINESE WAY" and is ideally suited for a low calorie, low cholesterol diet, that is both simple to follow and yet very tasty in its end result.

From personal experience of Mr. Yan's dishes I can wholeheartedly recommend his book of recipes as being both nutritious and satisfying to the palate.

Bon appetit on the road to low calories and better health.

*George. H. Changfoot*

George H. CHANGFOOT, CARDIOLOGIST,
New Westminster, B. C., Canada,
1978.

# WHY VEGETABLE COOKING?

**Because of the following three main reasons:**

## (1) HEALTH

— Vegetables are rich in vitamins, such as vitamins C and D. If vegetables are cooked in a short period of time, such as the Chinese stir-fry method, these vitamins can easily be absorbed in our system.

— Some vegetables, such as beans and Tofu, have as much nutritional value as meat to provide essential protein but contain no cholesterol and low calorie. Vegetarians generally have low incidence of high blood pressure and heart attacks. It is believed in China that celery has a hypotensive effect whilst ginger roots increase blood circulation after consumed.

— Most vegetables are less spicy, therefore present less irritation to the alimentary tract.

— When consumed regularly, the roughage stimulates peristalsis in the bowel and thus preventing constipation.

## (2) ECONOMY

— Vegetables are less expensive and are grown easily in the garden. They can be frozen in convenient packages.

## (3) VARIETY

— Vegetables are available in much greater variety than meats. They are more colourful and can be mixed in multitudes of servings.

# BASIC INGREDIENTS FREQUENTLY USED IN CHINESE COOKING

Chinese cooking ingredients and utensils are available at the YAN'S VARIETY COMPANY LTD., Post Office Box 227, Port Coquitlam, British Columbia, V3C 3V7, CANADA. Phone number 941-6607. Mail orders are accepted.

## CHINESE GOURMET POWDER
Monosodium Glutamate (M.S.G.), made from cereals in China. Salt-like in appearance, has the virtue of bringing out and accentuating the flavour of any food with which it is employed.

## CHINESE COOKING WINE
Special rice wine has the effectiveness in the marination of meats particularly in chicken. It is a meat tenderizer and flavour promoter.

## DRIED MUSHROOMS
Black mushrooms are sold in dried state. Can be kept for a long time. Soaking in hot water for half an hour is necessary before use. Produce exotic flavour and smell in Chinese dishes.

## GINGER
Knotty knobs of ginger roots are used frequently in Chinese cooking. It can be kept for weeks in a cool dry place. Refrigeration is not required. It should be peeled before use. Powdered ginger can be used as a substitute.

## FIVE SPICE POWDER
A special mixture of 5 spices from China. Very strong in flavour and is an appetizing agent in cooking. Used mainly for meat dishes.

## HOI SIN SAUCE
A highly seasoned sauce, brown in colour, made from Chinese pumpkin, sugar, spices and soy sauce. Used in barbecuing, steaming and stir frying.

## OYSTER SAUCE
A thick brown sauce with strong oyster taste. Sold in bottles or cans. Can be kept for a long time without refrigeration. A delicious sauce.

# TAPIOCA STARCH POWDER

Very fine white powder similar to cornstarch but more starchy and turns any food from a dull colour to an attractive glossy look. Used frequently to thicken sauce or gravy. It has a mild tenderizing effect on meats, therefore it is used in various marinations.

# SESAME OIL

Reddish-brown oil has the effect to add flavour to any kind of dishes. To be used in terms of drops only. It has a strong smell of sesame, sold in small bottles and does not require refrigeration.

# SALTED BLACK BEANS

Soft, fermented black beans sold in plastic bags. An ingredient in Chinese cooking that produces an exotic and delicious flavour. They can be kept indefinitely in a covered container if stored in a refrigerator.

# LIGHT SOY SAUCE

An important item in Chinese cooking as salt is to American cookery. It adds distinctive flavour in marination. Light brown in colour, salty but does not stain vegetables in cooking.

# DARK SOY SAUCE

Usually found in Chinese specialty shops. Dark brown in colour, has a sweet taste and is used frequently for stewing and food colouring in Chinese cooking. Can be kept for years without refrigeration.

# CHINESE PEANUT OIL

Pure 100 per cent peanut with a delightful smell and flavour to make Chinese cooking more delicious and appetizing.

# YAN'S ALMIGHTY POWDER

An all purpose battermix powder prepared with egg white powder, bread crumbs, flour, starch and salt. Just add equal amounts of water and powder, then mix into a smooth consistency. Excellent for all types of deep frying.

# TIPS AND HINTS FOR
# SUCCESSFUL CHINESE COOKING

## PLANNING:

— Consider your own ability and do not plan for more dishes than you can handle.
— Choose most dishes that can be cooked ahead of time and pick only one last minute dish for your menu.

## PREPARATION:

— Check the ingredient list for completeness. Assemble all the ingredients in one place, preferrably on a large tray near the stove for cooking.
— Preparation takes up more time than the actual cooking. Allow time to prepare all ingredients.
— Washing and cutting should be done ahead of time.
— Most ingredients are cut to uniform bite size, shape and thickness. Cutting diagonally increases the exposure to heat and ensures better cooking.
— Meats should be thinly sliced across grain to give more tenderness.
— Oil and seasoning ingredients should be readily accessible on counter.
— Marination should be done ahead of time.

## COOKING:

— Study and understand the procedure before any cooking.
— The skillet or wok must be hot and use high heat to do cooking unless specified.
— Prepare rice or noodles dishes first; cook vegetable dishes last.
— When food is being cooked with a lid, DO NOT take lid off until the time is up. If steam pressure is lost, the food will take longer time to cook.
— Cook quickly; serve immediately.

# PLANNING OF MENUS

Assuming 4 to 6 people in a family, a sufficient menu should consist of 1 soup, 1 rice, 1 dish with vegetables, and 3 meat dishes. For every addition of 2 persons, another meat dish should be added.

The contrast and diversity of dishes must be considered. When one is spicy, another one should be bland; if one is meaty, the other should be more vegetables; when there is 1 dish of chicken, the other should consist of other kinds of meat to avoid duplication; when filling dish like steamed rice is planned, one should avoid another filling dish like chow meins or fried rice.

Be honest with your own ability and do not plan several last minute dishes for your meal. It is always recommended that only one last minute stir-fry dish per menu be considered, and pick the rest that can be done ahead of time.

# TABLE OF CONTENTS

# ALMOND JELLO WITH FRUIT COCKTAIL

## INGREDIENTS:                    AMOUNT:

| Japanese Aga Aga (jello plant) | ½ ounce or 1 piece |
| Almond extract | 1 teaspoonful |
| Creamo | 2 cups |
| Sugar | ½ cup |
| Fruit cocktail | 1 can |

## METHOD:

(1)   Use a saucepan to boil up 4 cups of water. Dissolve jello plant and sugar in water and boil for 5 minutes.

(2)   Turn off the heat. Add creamo and almond extract. Stir well.

(3)   Use a sieve to strain liquid into a square cake pan.

(4)   Put into the freezing compartment of a refrigerator to set for 30 minutes.

(5)   When done, cut jelly into 1" cubes and pour into a large glass bowl.

(6)   Add fruit cocktail with juice on top of the jelly. Chill and serve in individual small bowls.

---

*"I watch your program every day
and if I could wok like you I would be okay.
My mouth waters with every dish,
especially when you wok the fish.
Please keep up the good wok
We enjoy your program — Thanks a lot!
At the end of each program you usually say
to write for the free recipes of the day.
This means with the dishes that you wok
I would be writing to you quite a lot."*

**Betty McLean, Toronto, Ontario, Canada**
\* \* \* \* \* \* \* \* \* \*

# ALMOND WITH MIXED VEGETABLES

| INGREDIENTS: | AMOUNT: |
|---|---|
| Raw Almond | 6 to 8 ounces |
| Onion | 1 medium, shredded |
| Fresh Mushrooms | 1 cup, cut into halves |
| Ginger Roots | 4 slices |
| Pepper | Dash |
| Sesame Seed Oil | Drops |
| Tapioca Starch | 1 tablespoonful |
| Light Soy Sauce | 1½ tablespoonful |
| Frozen Mixed Vegetables | 6 ounces, thawed |

## METHOD:

(1) Use hot peanut oil to deep fry almonds until golden brown. Keep warm in oven.

(2) In a hot wok, add 2 tablespoonful of peanut oil. When oil is hot, put in ginger, onion to brown for a minute. Then put in all the vegetables, ½ teaspoonful of salt, 2 tablespoonful of water. Cover with a lid. Cook with high heat for 3 minutes.

(3) Dissolve starch in ¼ cup of water, ½ tablespoonful of soy sauce, and put into the vegetable mixture to cook. Return almonds, mix and serve hot.

# APPLE FRITTERS

## INGREDIENTS:                    AMOUNT:

| Fresh apples | 3 medium size, peeled, cored and cut each into 6 wedges |
| Flour | 3 tablespoonful |
| Tapioca starch | 3 tablespoonful |
| Egg | 1 large, lightly beaten |
| Water | 1 tablespoonful |
| Syrup, hot | $2/3$ cup |
| Icy water | 1 bowl |
| Peanut oil | 16 ounces |

## METHOD:

(1) Coat a plate lightly with oil for later use.

(2) Use a saucepan to keep the syrup boiling on very low heat.

(3) Heat wok with 15 ounces of peanut oil. Use a bamboo chopstick to check the temperature of the oil. When bubbles appear around the stick in oil, it indicates the readiness for deepfrying.

(4) In a bowl, prepare batter mix with egg, flour, starch and water. Stir and mix until it becomes smooth in consistency. Dip apple in egg batter and deep fry until golden brown. Immediately coat apple in hot syrup and then dip fritters in icy water. Pick up immediately and place on plate. Serve at once.

# ASPARAGUS WITH FU-YUNG SAUCE

## INGREDIENTS:

Fresh asparagus

Peanut oil
Salt
YAN'S Fu-Yung Sauce, heated

## AMOUNT:

2 pounds, wash and break off
the tougher end
2 to 3 tablespoonful
½ teaspoonful
See page 48

## METHOD:

(1) Cut asparagus diagonally into 2 inch lengths.
(2) Use high heat to warm up wok, add peanut oil until smoke starts to rise. Put in asparagus, salt. Stir and mix well.
(3) Add 2 tablespoonful of water, cover with a lid. Cook with high heat for 3 minutes until steam escapes from the edge of lid. Remove and drain with a sieve or wire basket before putting on plate.
(4) Pour hot FU-YUNG SAUCE on top. Serve hot.

# ASPARAGUS WITH OYSTER SAUCE

## INGREDIENTS:      AMOUNT:

| INGREDIENTS: | AMOUNT: |
|---|---|
| Fresh Asparagus | 2 pounds |
| Peanut Oil | 2-3 tablespoonful |
| Salt | ½ teaspoonful |
| Oyster Sauce | 6 tablespoonful |
| Sugar | 1 teaspoonful |
| Tapioca Starch | 1 teaspoonful |
| Water | ¼ cup |
| Sesame Seed Oil | Drops |
| Light Soy Sauce | 1 teaspoonful |

## METHOD:

(1) Wash asparagus and drain in a sieve. Cut off the tougher end. Cut diagonally into 2 inch lengths.

(2) In a heated wok, add peanut oil until smoke starts to rise, put in asparagus, salt. Stir and mix well.

(3) Put in 2 tablespoonful of water, cover with a lid. Cook with high heat for 3-4 minutes until steam escapes from the edge of lid. Remove and drain with a sieve before putting on plate.

(4) Use a saucepan to boil up solution of oyster sauce, sugar, tapioca starch, water, sesame seed oil and light soy sauce. When done, apply over asparagus. Serve hot.

# ASPARAGUS WITH CRAB MEAT SAUCE

| INGREDIENTS: | AMOUNT: |
|---|---|
| Fresh Asparagus | 1 pound, cut at the middle into two |
| Crab meat | 4 to 6 ounces |
| Gourmet Powder | 1/8 teaspoonful |
| Pepper | Dash |
| Sesame Seed Oil | Drops |
| Salt | ¼ teaspoonful |
| Milk | 4 ounces |
| Water | 4 ounces |
| Tapioca Starch | 1 tablespoonful |
| Chicken Cube | 1 piece |

## METHOD:

(1) Use a saucepan to boil about 8 cups of water. Put in ½ teaspoonful of salt and 1 teaspoonful of oil.

(2) Put in asparagus, cover with a lid. Boil for 3 minutes. When done, remove to a plate and drain.

(3) Use a wok with high heat, dissolve chicken cube in 4 ounces of milk and 4 ounces of water. Put in gourmet powder, crab meat, pepper, sesame seed oil, salt. Boil for a minute.

(4) Put in starch solution to thicken the sauce and pour over the asparagus. Serve hot.

# BANANA FRITTERS

## INGREDIENTS:                    ## AMOUNT:

| | |
|---|---|
| Bananas, not too ripe | 3 medium, peeled, cross-cut diagonally into bite-size pieces |
| Flour | 3 tablespoonful |
| Tapioca starch | 3 tablespoonful |
| Egg | 1 large, lightly beaten |
| Water | 1 tablespoonful |
| Syrup | $2/3$ cup, keep boiling with low heat |
| Icy cold water | 1 cup |
| Peanut oil for deep fry | 15 ounces |

## METHOD:

(1)   Coat a plate with oil for later use.
(2)   Use a saucepan to keep the syrup boiling on very low heat.
(3)   Heat wok with peanut oil until ready for deep-fry.
(4)   In a bowl, stirmix flour, starch, egg and 1 tablespoonful of water to prepare a smooth batter-mix.
(5)   Dip banana in egg battermix and deep-fry until golden brown.
(6)   Coat banana in hot syrup and then dip in ice cold water for 1 second only. Immediately place on plate. Serve hot.

# BEAN SPROUTS
# WITH MIXED VEGETABLES

| INGREDIENTS: | AMOUNT: |
|---|---|
| Bean Sprouts, fresh | 1 pound |
| Celery | 2  stalks, sliced diagonnally into bite size pieces |
| Carrots | 1, sliced diagonally into bite size pieces |
| Broccoli | 1 piece, sliced diagonally into bite size pieces |
| Bamboo Shoots, sliced | 1 cup |
| Water Chestnuts, sliced | 1 cup |
| Ginger roots, minced | 1 tablespoonful |
| Salt | ½ teaspoonful |
| Starch Solution | 1 tablespoonful Tapioca Starch, ¼ cup water, drops of Sesame Seed Oil, 1 teaspoonful Light Soy Sauce |

## METHOD:

(1) Heat 2 tablespoonful peanut oil to very hot in wok.

(2) Put in ginger to brown for ½ minute. Add all vegetables, salt, 1 tablespoonful of water, mix, cover with a lid. Cook on a high heat for 3-5 minutes.

(3) When done, add starch solution and bring to a boil. Serve hot.

# BOK CHOY, STIR FRIED

## INGREDIENTS:

| INGREDIENTS: | AMOUNT: |
|---|---|
| Bok Choy | ¾ to 1 pound, slice white part of vegetable diagonally and cut the green leaves into 2 inch lengths |
| Ginger Root | 4 slices |
| Sugar | ½ teaspoonful |
| Salt | ½ teaspoonful |
| Peanut Oil | 2 tablespoonful |
| Light Soy Sauce | 1 teaspoonful |
| Tapioca Starch | 1 tablespoonful |
| Cooking Wine | ½ teaspoonful |

## METHOD:

(1) Use high heat to heat up wok; add peanut oil, ginger slices and stir fry for ½ minute.

(2) Put in white part of the Bok Choy and stir fry for 1 minute.

(3) Add the green leaves into the wok, salt, sugar, cooking wine, and 2 tablespoonful of water.

(4) Cover with a lid. Cook at a high heat for 3 - 4 minutes until steam escapes from the edge of the lid. When the vegetables are done, dissolve starch solution with 1 tablespoonful tapioca starch, 1 teaspoonful light soy sauce, few drops of sesame seed oil, if preferred, and ¼ cup of water. Stir well, add to the vegetables and bring to a boil. Serve hot.

# BROCCOLI, STIR FRIED

## INGREDIENTS:

| INGREDIENTS: | AMOUNT: |
|---|---|
| Broccoli | 3-4 stalks |
| Ginger Roots | 4-6 slices |
| Chinese Cooking Wine | 1 teaspoonful |
| Sugar | 1 teaspoonful |
| Salt | ½ teaspoonful |
| Peanut Oil | 2 tablespoonful |
| Tapioca Starch | 1 tablespoonful |
| Sesame Seed Oil | Few drops |
| Starch Solution | 1 tablespoonful Tapioca Starch, ¼ cup water, drops of Sesame Seed Oil, 1 teaspoonful Light Soy Sauce |

## METHOD:

(1) Cut tops of broccoli into flowerets, slice stalks into match sticks about 3 inches in length.

(2) In a heated wok, add peanut oil until smoke starts to rise.

(3) Put in ginger to brown for ½ minute. Then add broccoli, salt, sugar, wine and 2 tablespoonful water. Cover with a lid. Cook with a high heat for 3 - 5 minutes.

(4) Add starch solution and bring to a boil. Serve hot.

# BROCCOLI IN OYSTER SAUCE

| INGREDIENTS: | AMOUNT: |
|---|---|
| Broccoli, fresh | 3 medium stalks |
| Peanut Oil | 2-3 tablespoonful |
| Salt | $\frac{1}{2}$ teaspoonful |
| Oyster Sauce | 6 tablespoonful |
| Sugar | 1 teaspoonful |
| Tapioca Starch | 1 teaspoonful |
| Water | $\frac{1}{4}$ cup |
| Sesame Seed Oil | Drops |
| Light Soy Sauce | 1 teaspoonful |

## METHOD:

(1) Wash broccoli and drain in a sieve. Cut the tops into flowerets. Make stalk into match sticks about 3 inches in length.

(2) In a heated wok, add peanut oil until smoke starts to rise, put in broccoli, salt. Stir and mix well.

(3) Put in 2 tablespoonful of water, cover with a lid. Cook with a high heat for 3-4 minutes until steam escapes from the edge of lid. Remove and drain with a sieve before putting on plate.

(4) Use a saucepan to boil up solution of oyster sauce, tapioca starch, water, sesame seed oil and light soy sauce. When done, apply over broccoli. Serve hot.

# BROCCOLI WITH FU-YUNG SAUCE

| INGREDIENTS: | AMOUNT: |
| --- | --- |
| Fresh broccoli | 3 medium stalks |
| Peanut oil | 2 to 3 tablespoonful |
| Salt | ½ teaspoonful |
| YAN'S Fu-Yung Sauce, heated | See page 48 |

## METHOD:

(1) Wash broccoli and drain in a sieve. Cut the tops into flowerets. Make stalk into match sticks about 3 inches in length.

(2) In a heated wok, add peanut oil until smoke starts to rise, put in broccoli, salt. Stir and mix well.

(3) Put in 2 tablespoonful of water, cover with a lid. Cook with a high heat for 3-4 minutes until steam escapes from the edge of lid. Remove and drain with a sieve before putting on plate.

(4) Apply Fu-Yung Sauce. Serve hot.

# BRUSSELS SPROUTS, STIR-FRIED

## INGREDIENTS:          AMOUNT:

| INGREDIENTS: | AMOUNT: |
|---|---|
| Brussels Sprouts | 1 pound, cut into halves |
| Carrots | 1, sliced diagonally |
| Ginger Roots | 6 slices |
| Salt | ½ teaspoonful |
| Peanut Oil | 2 tablespoonful |
| YAN'S Fu-Yung Sauce, heated | See page 48 |

## METHOD:

(1) Heat 2 tablespoonful peanut oil to very hot in wok.
(2) Put in ginger to brown for ½ minute.
(3) Add vegetables, salt and 2 tablespoonful water, mix. Cover with a lid, cook with a high heat for 5 minutes.
(4) Remove to a plate. Add Fu-Yung Sauce. Serve hot.

# CAULIFLOWER WITH FU-YUNG SAUCE

| INGREDIENTS: | AMOUNT: |
|---|---|
| Cauliflower, fresh or frozen | 1 pound, cut into flowerets |
| Ginger roots, minced | 1 teaspoonful |
| Salt | ½ teaspoonful |
| Peanut Oil | 2 tablespoonful |
| YAN'S Fu-Yung Sauce, heated | See page 48 |

## METHOD:

(1)  Heat 2 tablespoonful peanut oil to very hot in wok.
(2)  Put in ginger to brown for ½ minute.
(3)  Add cauliflower, salt and 2 tablespoonful water, mix. Cover with a lid, cook with a high heat for 5 minutes.
(4)  Remove to a plate. Add Fu-Yung Sauce. Serve hot.

# CHINESE CABBAGE, STIR-FRIED

## INGREDIENTS:

| INGREDIENTS: | AMOUNT: |
|---|---|
| Chinese Cabbage (Siu Choy) | 1 pound, cut into 1 inch pieces |
| Ginger roots, minced | 1 tablespoonful |
| Salt | $\frac{1}{2}$ teaspoonful |
| Onion, shredded | 1 cup |
| Garlic, minced | 1 clove |
| Tapioca Starch | 1 tablespoonful |
| Light Soy Sauce | 1 teaspoonful |
| Sugar | $\frac{1}{2}$ teaspoonful |
| Peanut Oil | 2 tablespoonful |
| Sesame Seed Oil | Drops |

## METHOD:

(1) Use hot wok, add 2 tablespoonful peanut oil.

(2) Put in ginger, onion and garlic to brown on high heat for $\frac{1}{2}$ minute.

(3) Add cabbage, salt to mix. Then add 2 tablespoonful water, cover with lid. Cook on a high heat for 4-5 minutes.

(4) Dissolve starch in $\frac{1}{4}$ cup water. Add soy sauce, sesame seed oil and sugar. Add to vegetables. Bring to a boil. Mix and serve hot.

# CHINESE HOT TOSSED SALAD

| INGREDIENTS: | AMOUNT: |
|---|---|
| Fresh Cucumber | 1 medium, remove seeds and make into match-sticks about 3 inches in length |
| Carrot | 1 medium, matchsticked as above |
| Green Onions | 4 stalks, cut into 3 inch lengths |
| Bamboo Shoots | 1/2 cup, sliced thin |
| Green Pepper | 1 medium, slivered or match-sticked |
| Red Pepper | 1 medium, slivered or match-sticked |
| Chinese Cooking Wine | 1 tablespoonful |
| Sesame Oil | 1 teaspoonful |
| Vinegar | 1 tablespoonful |
| Sugar | 2 tablespoonful |
| Tobasco Sauce | 1 teaspoonful |
| Light Soy Sauce | 2 tablespoonful |
| Garlic | 2 cloves, minced |
| Ginger Root | 1 teaspoonful, minced |
| Celery | 2 stalks, matchsticked into 3 inch lengths |

## METHOD:

(1) Use a bowl, blend into a smooth sauce with light soy sauce, tobasco sauce, wine, vinegar, sugar, sesame oil.

(2) Use high heat and a hot wok, add 2 tablespoonful of peanut oil. When smoke starts to rise, add ginger and garlic to brown for 30 seconds, then all vegetables and 1/2 teaspoonful of salt. Stir constantly for 3 minutes. Then pour in the hot sauce and stir mix for another 2 minutes. Then serve.

# CHINESE TURNIP (LO PAK)

## INGREDIENTS:

| | AMOUNT: |
|---|---|
| Chinese Turnip | 1 pound piece, cut into ½ inch bite size pieces |
| Hoi Sin Sauce | 2 tablespoonful |
| Ginger Roots, minced | 1 tablespoonful |
| Sugar | 1 teaspoonful |
| Salt | ¾ teaspoonful |
| Green Onion | 2 stalks, cut into 2" lengths |
| Garlic, minced | 1 tablespoonful |
| Chinese Cooking Wine | 1 teaspoonful |
| Sesame Oil | Few drops |
| Peanut Oil | 2 tablespoonful |
| Starch Solution: | 1 tablespoonful Tapioca Starch, ¼ cup water, drops of sesame seed oil, 1 teaspoonful Light Soy Sauce |

## METHOD:

(1) Heat 2 tablespoonful peanut oil to very hot in wok.
(2) Put in ginger, garlic, Hoi Sin Sauce and stir fry for ½ minute.
(3) Add turnip slices, salt, sugar, wine, sesame seed oil. Mix well. Then add 2 tablespoonful water. Cook with lid on for 4 minutes at a high heat.
(4) Put in green onions and serve hot.

# CHINESE SALAD

## INGREDIENTS: | AMOUNT:

| INGREDIENTS: | AMOUNT: |
|---|---|
| Broccoli | 1, cut top in flowerets and stalk into match sticks 3 inches in length |
| Bean Sprouts, fresh | ½ pound |
| Celery | 2 stalks, cut into 2 inch match sticks |
| Peanuts, roasted | ¼ cup, chopped fine |
| Green Onion | 2 stalks, chopped fine |
| Baking Soda | ¼ teaspoonful |
| Peanut Oil | 1 tablespoonful |

## DRESSING:

| | |
|---|---|
| Light Soy Sauce | 1 tablespoonful |
| Vinegar | 1 tablespoonful |
| Sugar | 2 tablespoonful |
| Salt | ½ teaspoonful |
| Sesame Seed Oil | Few Drops |

## METHOD:

(1) Use wok to boil 8 cups of water.
(2) Add baking soda, salt, peanut oil 1 tablespoonful. Then add broccoli, celery, bean sprouts. Stir mix ½ minute. Remove immediately and rinse under running cold water.
(3) Drain water off and put in refrigerator to chill for 1 hour.
(4) Mix dressing ingredients, pour onto vegetable salad and mix well.
(5) Garnish with green onions and peanuts.

# EGG DROP SOUP

## INGREDIENTS:

## AMOUNT:

| | |
|---|---|
| Eggs | 2 medium, beaten with dash of salt |
| Gourmet powder | 1/4 teaspoonful |
| Salt | 1/2 teaspoonful |
| Green onion | 2 stalks, chopped fine |
| Soup stock | 6 cups |
| Pepper | Dash |
| Green peas | 1 cup |

## METHOD:

(1) Prepare soup stock in a saucepan and bring to a boil with high heat.

(2) Add pepper, salt and green peas into soup and boil for a minute.

(3) Slowly pour the beaten eggs in a thin stream. Stir constantly and slowly in a circular fashion until the eggs form thin shreds in the hot soup.

(4) Serve in individual soup bowls. Garnish with chopped green onions before serving.

*"Seeing your program is a*
*"wokful" delight*
*I'm eating Chinese food*
*every night.*
*Please send me your cookbook*
*"on the dot"*
*I need more, ideas to*
*keep on the "wok"!"*

*Gilles Ferland, Vensun, Quebec, Canada*
* * * * * * * * * *

33

# EGG FU-YUNG

## INGREDIENTS:

| INGREDIENTS: | AMOUNT: |
|---|---|
| Cabbage | 1 cup, shredded |
| Pepper | Dash |
| Eggs | 5 eggs, beaten with pinch of salt |
| Bean Sprouts | 1 cup |
| Green Pepper | 1 small, shredded |
| Onion | 1 small, shredded |
| Green Onion | 1 stalk, chopped fine |
| Salt | ¼ teaspoonful |
| Tapioca Starch | 2 tablespoonful |
| Gourmet Powder | Pinch |
| Light Soy Sauce | 1 teaspoonful |
| Sesame Seed Oil | Drops |
| YAN'S Fu-Yung Sauce | See page 48 |

## METHOD:

(1) Use high heat and 2 tablespoonful of peanut oil, put in onion, cabbage, bean sprouts, gourmet powder, salt, green pepper. Stir fry for 2 minutes. Remove from wok and let cool in a large bowl.

(2) Put in eggs, green onions with the vegetable mixture. Stir until well mixed.

(3) Use medium-high heat, 1 tablespoonful of peanut oil, add ½ cup of egg mixture into wok. Cook for 3 minutes until brown. Turn over like a pancake and cook the other side. Repeat same procedure for the rest of mixture. Serve with Fu-Yung Sauce.

# EGG WITH BEAN SPROUTS

| INGREDIENTS: | AMOUNT: |
|---|---|
| Eggs | 2 large, beaten with dash of salt |
| Bean Sprouts, fresh | 1 pound, rinse and drain |
| Celery | 2 stalks, slice diagonally |
| Mushrooms | 2 - 4 ounces, cut into slices |
| Onion | 1 medium, cut into wedges |
| Garlic | 1 clove, minced |
| Ginger | 1 tablespoonful, minced |
| Green Onion | 3 stalks, cut into 2 inch lengths |
| Salt | ½ teaspoonful |
| Tapioca Starch | 1 tablespoonful |
| Light Soy Sauce | 1 teaspoonful |
| Sesame Seed Oil | Few drops |
| Sugar | ½ teaspoonful |

## METHOD:

(1) Use high heat in the wok, add 1 tablespoonful of peanut oil until smoke begins to rise.

(2) Pour beaten egg into wok. Turn around sideways to make the egg go bigger. When the egg becomes almost solid use spatula to turn egg over and cook egg on the other side for another minute. When done, remove to a chopping block and clean the wok.

(3) In hot wok use 2 tablespoonful of peanut oil and when the smoke begins to rise put in garlic and ginger for ½ minute.

(4) Add all vegetables, salt, and stir mix for ½ minute. Add 1 tablespoonful water and cover with a lid to cook with high heat for 3 minutes.

(5) While waiting, use a cleaver to make fried egg into ½ inch slivers. Put on a plate for later use.

(6) Prepare starch solution with tapioca starch, light soy sauce, sesame seed oil, sugar and ¼ cup water.

(7) When vegetables are done pour in starch solution and bring to a boil. Add egg and mix well. Serve immediately.

# EGGPLANT, DEEP FRIED

| INGREDIENTS: | AMOUNT: |
|---|---|
| Eggplant | 1 medium, cut into ¼ inch slices |
| Salt | 1 tablespoonful |
| Flour | 1 cup |
| Chinese 5 Spice Powder | ¼ teaspoonful |
| Egg | 1, lightly beaten |
| Baking Powder | ¼ teaspoonful |
| Pepper | Pinch |
| Peanut Oil | 15 ounces |

## METHOD:

(1) Sprinkle eggplant with salt, 5 spice powder. Allow to stand for 20 minutes.
(2) Prepare batter mix in a bowl with flour, baking powder, egg and ¾ cup water until pasty in consistency.
(3) Dip eggplant slices in batter mix. Deep fry in hot peanut oil for 5 minutes until golden brown. Serve with salt and pepper.

---

*"...I think you are one of the funniest cooks to every cook such master pieces..."*

*Anthony Moretti, Frankfort, N.Y., U.S.A.*
\* \* \* \* \* \* \* \* \* \*

---

*"...I am eight years old and am home sick from school, I really like your show..."*

*Nerissa Davies, Comox, B.C., Canada*
\* \* \* \* \* \* \* \* \* \*

# EGGPLANT IN BLACK BEAN SAUCE

## INGREDIENTS:

| | AMOUNT: |
|---|---|
| Eggplant | 1 pound, sliced diagonally in $\frac{1}{2}$ inch lengths |
| Salted Black Beans | 3 tablespoonful, rinsed and drained |
| Garlic, minced | 1 tablespoonful |
| Ginger, minced | 1 tablespoonful |
| Gourmet Powder | $\frac{1}{4}$ teaspoonful |
| Chinese Cooking Wine | 1 tablespoonful |
| Sugar | 1 teaspoonful |
| Green Onion, chopped | $\frac{1}{2}$ cup |
| Onion, shredded | 1 cup |
| Salt | $\frac{1}{2}$ teaspoonful |
| Starch Solution | 1 tablespoonful Tapioca Starch $\frac{1}{4}$ cup water, drops of Sesame Seed Oil, 1 teaspoonful Light Soy Sauce |

## METHOD:

(1)  Use a bowl to hold black beans, minced ginger, garlic, and crush these ingredients into a paste. Then add soy sauce, sugar, wine, gourmet powder, sesame oil. Mix well with 2 tablespoonful of water.

(2)  Use 2 tablespoonful of peanut oil in very hot wok. Put in onion, black bean paste. Stir fry $\frac{1}{2}$ minute.

(3)  Add eggplant, salt, $\frac{1}{4}$ cup water and mix well. Cover with lid. Cook with a high heat for 3-5 minutes.

(4)  When done add starch solution and bring to a boil. Serve hot. Garnish with chopped green onion.

# EGGPLANT, SPICED

| INGREDIENTS: | AMOUNT: |
|---|---|
| Eggplant | 1 medium, cut in half, then sliced diagonally into ½ inch strips |
| Garlic, minced | 1 tablespoonful |
| Ginger roots, minced | 1 tablespoonful |
| Onion, shredded | 1 cup |
| Hoi Sin Sauce | 2 tablespoonful |
| Celery, diced | 1 cup |
| Salt | ½ teaspoonful |
| Starch Solution | 1 tablespoonful Tapioca Starch, ¼ cup water, drops of Sesame Seed Oil, 1 teaspoonful Light Soy Sauce |

## METHOD:

(1) Heat 2 tablespoonful of peanut oil to very hot in wok.

(2) Put in onion, ginger, garlic to brown for ½ minute. Put in Hoi Sin Sauce and stir mix, then eggplant slices, celery, salt, and 2 tablespoonful water. Mix well. Cover with a lid. Cook at a high heat for 3 minutes.

(3) When done add starch solution, bring to a boil and serve hot.

# GREEN BEANS WITH HOI SIN SAUCE

## INGREDIENTS:                    AMOUNT:

| Green Beans | 1 pound, cut into 2" length |
| Hoi Sin Sauce | 3 tablespoonful |
| Light Soy Sauce | 1 tablespoonful |
| Tapioca Starch | 1 tablespoonful |
| Garlic | 2 cloves, crushed and skinned |
| Onion | 1 medium, cut into sixths |
| Gourmet Powder | 1/8 teaspoonful |
| Chinese Cooking Wine | 1 teaspoonful |
| Sesame Seed Oil | Drops |
| Sugar | 1/2 teaspoonful |

## METHOD:

(1)  Use high heat and a hot wok, put in 2 tablespoonful of peanut oil. When hot, add garlic, onion and Hoi Sin Sauce to brown for a minute. Then put in green beans, 1/4 teaspoonful of salt, wine, sugar, gourmet powder, 1/4 cup of water. Cover with a lid and cook with a high heat for 3-5 minutes.

(2)  When done, put in starch solution and bring to a boil. Mix and serve hot.

---

*"There is a young man named Yan*
*He is not just a flash in the pan*
*How he loves to talk while he woks with his wok*
*He stirs it up quick with his little chop stick*
*It is all cooked up fresh and not from a can*
*And we like your show, Mr. Yan!!!"*

*Mrs. Shirley Engle, Alderwood Manor, Washington, U.S.A.*
* * * * * * * * *

# MUSHROOM AND SPINACH SOUP

| INGREDIENTS: | AMOUNT: |
|---|---|
| Fresh Spinach | 6 ounces, cut into 2 inch sections |
| Mushrooms, fresh or canned | 1 cup |
| Salt | 1 teaspoonful |
| Soup Stock | 8 cups |
| Gourmet Powder | ¼ teaspoonful |
| Garlic | 1 clove, minced |
| Peanut Oil | 1 teaspoonful |

## METHOD:

(1) Use high heat on a wok. Add 1 teaspoonful of peanut oil. When hot put in garlic to brown for 5 seconds, then mushrooms and spinach. Stir mix for 1 minute.

(2) Add soup stock, salt, gourmet powder and boil for 2 minutes. Serve hot.

# MUSHROOMS, CRISPY

## INGREDIENTS:

| INGREDIENTS: | AMOUNT: |
|---|---|
| Fresh Mushrooms | 1 to 2 pounds, washed and drained |
| Light Soy Sauce | 2 tablespoonful |
| Sugar | 1 tablespoonful |
| Chinese Cooking Wine | 2 tablespoonful |
| Gourmet Powder | $\frac{1}{4}$ teaspoonful |
| Salt | $\frac{1}{2}$ teaspoonful |
| Pepper | Dash |
| Sesame Oil | Drops |

## METHOD:

(1)  Marinate mushrooms for 30 minutes with light soy sauce, sugar, gourmet powder, salt, pepper, sesame oil and cooking wine.

(2)  Prepare batter-mix as listed on page 17, or dissolve YAN'S ALMIGHTY POWDER with equal amount of water to make a smooth pasty consistency.

(3)  Dip mushrooms in batter-mix and deep fry in hot peanut oil until golden brown. Remove with wire basket, drain on paper towel and serve hot.

N.B. Can be served with plum sauce or oyster sauce on tooth picks.

# MUSHROOMS IN OYSTER SAUCE

## INGREDIENTS:

| INGREDIENTS: | AMOUNT: |
|---|---|
| Mushrooms, fresh | 1 pound, cut into halves |
| Green Onion | 2 stalks, cut into 2 inch lengths |
| Ginger Roots | 1 tablespoonful, minced |
| Chinese Cooking Wine | 1 teaspoonful |
| Sugar | 1 teaspoonful |
| Oyster Sauce | 3 tablespoonful |
| Dark Soy Sauce | 1 teaspoonful |
| Tapioca Starch | 1 tablespoonful |
| Gourmet Powder | $\frac{1}{4}$ teaspoonful |
| Sesame Seed Oil | Few Drops |

## METHOD:

(1) Use a high heat on a wok, add 2 tablespoonful peanut oil to brown garlic, ginger and onion for $\frac{1}{2}$ minute.

(2) Add mushrooms to stir fry for $\frac{1}{2}$ minute. Then add salt, cooking wine, 2 tablespoonful of water.

(3) While waiting, prepare oyster sauce with tapioca starch, oyster sauce, sugar, water, sesame seed oil, dark soy sauce and gourmet powder in a bowl.

(4) When mushrooms are done, add oyster sauce into the wok and bring to a boil. Mix and serve immediately.

# MUSHROOMS WITH BAMBOO SHOOTS

## INGREDIENTS:

| INGREDIENTS: | AMOUNT: |
|---|---|
| Mushrooms, fresh or canned | ½ pound |
| Bamboo Shoots, sliced | 5 - 6 ounces |
| Celery | 1 - 2 stalks, sliced diagonally |
| Carrot | 1, sliced diagonally |
| Snow Peas, if available | 3 - 4 ounces |
| Salt | ½ teaspoonful |
| Sugar | ½ teaspoonful |
| Sesame Seed Oil | Few drops |
| Tapioca Starch | 1 tablespoonful |
| Light Soy Sauce | 1 teaspoonful |
| Oyster Sauce | 2 tablespoonful |
| Garlic | 2 cloves, minced |

## METHOD:

(1) Use high heat and hot wok, add 2 tablespoonful peanut oil. When the smoke begins to rise add garlic to brown for ½ minute. Then put in all vegetables, salt, and stir mix for ½ minute.

(2) Add 2 tablespoonful of water. Cover with a lid. Cook with high heat for 3 minutes.

(3) While waiting, prepare tapioca starch solution with 1 tablespoonful tapioca starch, ¼ cup water, drops of sesame seed oil, light soy sauce, oyster sauce and sugar.

(4) When the vegetables are done stir starch solution in and bring to a boil. Serve hot.

# PEAS AND MUSHROOMS

## INGREDIENTS:                    AMOUNT:

| INGREDIENTS: | AMOUNT: |
|---|---|
| Peas, fresh or frozen | 10 ounces |
| Bamboo Shoots, sliced | ½ cup |
| Mushrooms, fresh, canned or frozen | 1 cup |
| Celery, diced | 1 cup |
| Salt | ¾ teaspoonful |
| Onion, shredded | ½ cup |
| Garlic, minced | 1 tablespoonful |
| Tapioca Starch | 1 tablespoonful |
| Sesame Oil | Drops |

## METHOD:

(1)  Heat 2 tablespoonful peanut oil to very hot in wok.

(2)  Put in garlic, onion to brown for ½ minute.

(3)  Add all vegetables, mix. Then put in salt and 2 tablespoonful water. Cover with lid and cook at a high heat for 3-5 minutes.

(4)  Add starch solution (prepared with starch, ¼ cup water, drops of sesame seed oil) and bring to a boil.

---

*"...first let me tell you how much my husband & I enjoy your cooking. Woking with you is something we look forward to every day, in fact we have invited some of our friends to join us, as they can't get your station..."*

*Madeline Reimer, Dolgeville, N.Y., U.S.A.*
\* \* \* \* \* \* \* \* \* \*

# STEAMED RICE

*A guaranteed and fool-proof method to cook steamed rice. Long-grain unpolished rice is used. It is less gummy and very suitable for fried rice. 1 cup of uncooked rice yields 3 cups of cooked rice.*

## INGREDIENTS:          AMOUNT:

| INGREDIENTS | AMOUNT |
| --- | --- |
| Long grain rice | 1 cup |
| Water | $1\frac{1}{2}$ cup |

When more rice is needed, just add same proportional amount of water.

## METHOD:

(1)   Wash rice by rubbing rice between palms of hands. Drain off all water.

(2)   Add the correct amount of water into the rice. Any temperature of water can be used. Do not put salt or butter to cook rice as it will destroy the sweet flavour of good steamed rice.

(3)   Cook the rice with a saucepan over high heat. UNCOVERED, until tiny holes or craters formed over the surface of rice. Switch to low heat and cover closely with a lid. Simmer the rice for 15 to 20 minutes. DO NOT TAKE LID OFF until time is up. This is the most critical time as the rice is steam-cook under pressure. When done, stir and serve hot.

Left-over steamed rice can be saved and kept in refrigerator until sufficient for making fried rice. As a matter of fact, freshly cooked rice is too soft and not suitable for fried rice. Try to cook the rice ahead of time, if you need cooked rice for frying. Also fluff up the cooked rice before storing can make frying easier to handle and improve the result in a professional appearance.

# FRIED RICE

## INGREDIENTS:    AMOUNT:

| INGREDIENTS: | AMOUNT: |
|---|---|
| Cooked rice, cold | 4 to 6 cups |
| Egg | 2 large, beaten light with dash of salt |
| Salt | ½ teaspoonful |
| Chinese cooking wine | ½ teaspoonful |
| Dark soy sauce | 2 tablespoonful |
| Green onions | 2 stalks, chopped fine |
| Green peas, frozen | 1 cup, thawed |

## METHOD:

(1) With a hot wok and medium heat, add 1 tablespoonful of peanut oil. Put in the eggs to cook for few minutes until the egg becomes solid. Then turn over to cook the other side for another minute. When done, remove to a cutting board and cut into slivers.

(2) Use medium heat to high heat, add 3 tablespoonful of peanut oil.

(3) Put in cooked rice, salt, wine and dark soy sauce. Keep stirring until rice is hot.

(4) Add peas, cooked eggs and chopped green onions. Stir for another minute and serve hot.

# ONION CREPES

## INGREDIENTS:

| | AMOUNT: |
|---|---|
| Flour | 1 cup |
| Hot water | 1/3 cup |
| Green onions | 3 stalks, chopped fine |
| Sesame oil | 1/2 teaspoonful |
| Salt | 1/2 teaspoonful |
| Onion | 2 tablespoonful, chopped fine |

## METHOD:

(1) In a bowl, mix flour with hot water and knead into a dough for 5 minutes.

(2) Put in green onions, onions, salt and sesame oil and knead for another 5 minutes.

(3) Make into a long roll and then cut into 6 pieces.

(4) Use a rolling pin to make each piece into a 4 inch round crepe.

(5) Use a hot wok and medium heat, put in 2 tablespoonful of peanut oil to fry crepe on both sides until brown in colour. Can be preapred ahead of time and used as snacks.

---

*"...you're a cook after my own heart...I enjoy your program immensely! You're a great comedian as well as a great cook..."*

**Diane Stomtard, Toronto, Ontario, Canada**

\* \* \* \* \* \* \* \* \* \*

# SAUCES

## YAN'S FU-YUNG SAUCE

| INGREDIENTS: | AMOUNT: |
|---|---|
| Egg white | 2 |
| Milk | 2/3 cup |
| Tapioca Starch | 2 tablespoonful |
| Peanut Oil | 2 tablespoonful |
| Salt | 1/2 teaspoonful |
| Pepper | Dash |
| Gourmet Powder | 1/4 teaspoonful |
| Water or Broth | 2/3 cup |
| Sesame Oil | Drops |

### METHOD:

(1) Beat egg white and milk for 30 seconds.
(2) Dissolve tapioca starch in water or broth, then add sesame oil, peanut oil, gourmet powder, salt, pepper.
(3) Use a wok with medium heat, bring the above 2 mixtures to a boil. Stir constantly until the solution becomes smooth and cooked.

## SWEET AND SOUR SAUCE

| INGREDIENTS: | AMOUNT: |
|---|---|
| Tapioca Starch | 2 tablespoonful |
| Sugar | 1/2 cup |
| Vinegar | 1/3 cup |
| Tomato Paste | 4 tablespoonful |

### METHOD:

Dissolve starch in 3/4 cup water, add sugar, vinegar and paste. Stir and bring to a boil.

# SNOW PEAS WITH BABY CORNS

## INGREDIENTS:                AMOUNT:

| | |
|---|---|
| Snow Peas, fresh or frozen | 1 pound |
| Baby Corns | 1 can, drained |
| Water Chestnuts | 6 ounces, sliced |
| Celery, diced | 1 cup |
| Carrots, sliced | 1 cup |
| Ginger roots, minced | 1 tablespoonful |
| Salt | ¾ teaspoonful |
| Tapioca Starch | 1 tablespoonful |
| Sesame Seed Oil | Drops |
| Onion, shredded | ½ cup |

## METHOD:

(1)  Heat 2 tablespoonful peanut oil to very hot in wok.
(2)  Put in ginger to brown for ½ minute.
(3)  Add all vegetables, salt, 2 tablespoonful water. Mix well. Cover with a lid to cook with high heat for 3-5 minutes.
(4)  Add starch solution (prepared with starch, ¼ cup water, drops of sesame seed oil) and bring to a boil. Mix and serve hot.

---

*"...I never leave the house until your program is over and all my family and friends never call me while I'm watching you, because they know I won't answer the phone..."*

**Mamie DeAngelo, East Detroit, Michigan, U.S.A.**
\* \* \* \* \* \* \* \* \* \*

# SPINACH, QUICK-FRIED

## INGREDIENTS:          AMOUNT:

Fresh Spinach            2 pounds, washed thoroughly
                         and cut into 3 inch length,
                         drained in a sieve

Salt                     ½ teaspoonful
Light Soy Sauce          2 tablespoonful
Sugar                    ½ teaspoonful
Sesame Oil               Drops
Oyster Sauce             2 tablespoonful
Chinese Cooking Wine     ½ teaspoonful
Carrot                   1 cup, match-sticked
Tapioca Starch           2 teaspoonful
Bamboo Shoots            ½ cup, sliced
Garlic                   1 clove, minced

## METHOD:

(1) When the wok is hot, use high heat to heat up 2 tablespoonful of peanut oil until smoke starts to rise.

(2) Put in garlic to brown for 10 seconds. Then add all the vegetables, salt, cooking wine to fry for 5 minutes. Turn with spatula constantly.

(3) Dissolve starch in ¼ cup of water, put in sugar, oyster sauce, light soy sauce and sesame oil. Stir well. Add to the vegetables. Bring to a boil and serve hot.

# SPROUTS 'N' PEAS

## INGREDIENTS:                    AMOUNT:

| INGREDIENTS: | AMOUNT: |
|---|---|
| Bean Sprouts | 1/2 pound |
| Water Chestnuts, sliced | 1/2 cup |
| Snow Peas | 1/2 - 1 pound |
| Salt | 1/2 teaspoonful |
| Garlic, minced | 1 teaspoonful |
| Carrot | 1, sliced diagonally |
| Starch Solution | 1 tablespoonful Tapioca Starch, 1/2 cup water, drops of Sesame Seed Oil, 1 teaspoonful Light Soy Sauce |

## METHOD:

(1)   Heat 2 tablespoonful peanut oil to very hot in a wok.
(2)   Put in ginger to brown for 1/2 minute.
(3)   Add all vegetables, salt, 1 tablespoonful water, stir mix. Cover with a lid. Cook on a high heat for 3-5 minutes.
(4)   When done, add starch solution and bring to a boil. Serve hot.

---

*"I like your wok,
I like your talk.
Your cooking is incredible
I want to learn
Oh, how I yearn
To make my food more edible.
The oil you heat,
You slice the meat,
It's such a simple matter;
Then comes the fish,
And now I wish
To know what's in the batter!"*

*Hazel Loring, Birmingham, MI, U.S.A.*

\* \* \* \* \* \* \* \* \* \*

# SWEET POTATO BALLS

## INGREDIENTS:                    ## AMOUNT:

Sweet Potatoes or Yams          1 pound, skinned and cut into
                                slices
Sugar                           ¼ cup
Flour                           ½ cup
Tapioca Starch                  2 tablespoonful
Peanut Oil                      15 ounces, for deep frying

## METHOD:

(1)   Boil potatoes until cooked. Add sugar, mix and mash.
(2)   Add flour, starch and knead well.
(3)   Make 1 inch balls.
(4)   Deep fry in hot oil until golden brown. Serve hot.

# TOFU WITH OYSTER SAUCE

## INGREDIENTS:                    AMOUNT:

Tofu (Bean Curd Cake)      2 squares, cut into ½ inch
                                           slices
Oyster Sauce                    4 tablespoonful
Sugar                              1 teaspoonful
Tapioca Starch                 1 tablespoonful
Green Onion                     2 stalks, chopped
Sesame Seed Oil              Few Drops

## METHOD:

(1)   Put Tofu on paper towel to dry for 1 hour.
(2)   Deep fry Tofu in hot peanut oil until golden brown.
       When done, remove to a plate and keep warm in oven.
(3)   Dissolve starch in ¼ cup water and mix with oyster
       sauce, sugar and sesame seed oil.
(4)   Heat starch solution to a boil and pour on top of Tofu.
       Garnish with chopped green onion.

# TOFU WITH SPINACH SOUP

## INGREDIENTS:

## AMOUNT:

| | |
|---|---|
| Tofu (Bean Curd Cake) | 1 piece, cut into 1 inch cubes |
| Spinach | ½ pound, washed and cut into 2 inch lengths |
| Soup Stock | 8 cups |
| Pepper | Dash |
| Sesame Seed Oil | Few Drops |
| Gourmet Powder | ¼ teaspoonful |
| Peanut Oil | ½ teaspoonful |
| Salt | ½ teaspoonful |

## METHOD:

(1) Use a saucepan or wok to bring soup stock to a boil.

(2) Add peanut oil, gourmet powder and salt.

(3) Put in Tofu and spinach, Sprinkle pepper and light soy sauce to taste. Boil for 2 minutes, add sesame seed oil and serve hot.

# TOMATO EGG FLOWER SOUP

## INGREDIENTS:

| | AMOUNT: |
|---|---|
| Tomato | 1 large, sliced into 8 wedges |
| Eggs | 2, lightly beaten |
| Light Soy Sauce | 1 tablespoonful |
| Salt | 1 teaspoonful |
| Green Peas | ¼ cup |
| Green Onion | 1 stalk, chopped |
| Gourmet Powder | ½ teaspoonful |
| Tapioca Starch | 2 tablespoonful |

## METHOD:

(1)  Use a hot wok, add 1 tablespoonful of peanut oil, put in green onion and tomato to fry for 1 minute. Then add 6 cups of water, salt, soy sauce, gourmet powder, green peas. Bring to a boil.

(2)  Dissolve starch in 2 tablespoonful water.

(3)  Slowly pour eggs in a thin stream into soup, then starch solution. Serve hot.

# VEGETABLE BUNS

## INGREDIENTS:                    AMOUNT:

| | |
|---|---|
| Flour | 6 cups |
| Sugar | 1 tablespoonful |
| Yeast | 2 teaspoonful |
| Lard or Margarine | 2 tablespoonful |
| Water | 1 ¾ cups |
| Cabbage | 1 pound, shredded |
| Celery | 1 cup, shredded |
| Onion | 1 large, shredded |
| Garlic | 2 cloves, minced |
| Salt | 1 tablespoonful |
| Gourmet Powder | 1 teaspoonful |
| Sesame Oil | ½ teaspoonful |
| Peanut Oil | 3 tablespoonful |
| Green Onions | 4 stalks, chopped fine |

## METHOD:

(1) Mix sugar in water, then sprinkle yeast on top and let stand for 15 minutes.

(2) Add melted lard, sifted flour and knead until smooth. Place in a bowl. Cover with a damp cloth. Wait for 5 hours until the dough becomes triple in size.

(3) Use a hot wok, 3 tablespoonful of peanut oil to brown garlic and onions. Then add all vegetables, salt, gourmet powder, 1 tablespoonful of water. Cover with a lid. Cook at high heat for 3 minutes. When done, remove to a plate and let cool.

(4) Remove risen dough and knead on a floured board for 5 minutes until smooth. Make into a roll and cut into 20 pieces. Roll each into a ball and flatten into 4-5 inch round patties.

(5) Put 1-2 tablespoonful of vegetable filling in middle, pick up edges in small pleats to close like a dome.

(6) Place on a square of wax paper and allow to rise for 5 minutes.

(7) Steam for 10 minutes over high heat.

# VEGETABLE CHOP SUEY

## INGREDIENTS:                    AMOUNT:

| | |
|---|---|
| Salt | $\frac{1}{2}$ teaspoonful |
| Bean sprouts | $\frac{1}{2}$ pound |
| Cabbage | $\frac{1}{2}$ pound, cut into bite size pieces |
| Celery | 2 stalks, cut diagonally into bite size pieces |
| Green Pepper | 1, shredded |
| Bamboo Shoots | $\frac{1}{2}$ small can, sliced thinly |
| Waterchest Nuts | $\frac{1}{2}$ small can, cut into halves |
| Carrots | 1, sliced diagonally |
| Starch Solution | 1 tablespoonful Tapioca Starch, $\frac{1}{2}$ cup water, drops of Sesame Seed Oil, 1 teaspoonful Light Soy Sauce |

## METHOD:

(1) Bring wok to high heat, add 2 tablespoonful of peanut oil. When the oil is hot, put in all vegetables, $\frac{1}{2}$ teaspoonful salt, 2 tablespoonful water. Stir mix. Cover with a lid. Cook with high heat for 3-4 minutes until steam is coming out from the lid.

(2) Add starch solution. Bring to a boil and serve immediately.

# VEGETABLE CHOW MEIN, AMERICAN STYLE

## INGREDIENTS:          AMOUNT:

| | |
|---|---|
| Gourmet Powder | ¼ teaspoonful |
| Pepper | Dash |
| Light Soy Sauce | 2 tablespoonful |
| Sugar | ¼ teaspoonful |
| Fresh Bean Sprouts | ½ pound |
| Onion | 1 medium, shredded |
| Mushrooms | ½ cup, cut into sections |
| Celery | 1 stalk, cut diagonally into bite size pieces |
| Green Pepper | 1 small, slivered |
| Salt | ½ teaspoonful |
| Tapioca Starch | 1 tablespoonful |
| Ginger | 3 slices |
| Crisp Noodles | 1 package |
| Oyster Sauce | 2 tablespoonful |

## METHOD:

(1) Heat up crisp noodles in oven at 350 °F for 15 to 20 minutes.

(2) Use a bowl to prepare sauce with tapioca starch, ¾ cup water, 1 tablespoonful light soy sauce, gourmet powder, oyster sauce and sugar.

(3) Use 3 tablespoonful of oil in hot wok, brown ginger, onion.

(4) Add vegetables, salt, ¼ cup water. Stir mix and cover with a lid. Cook with high heat for 3 minutes. Pour sauce in and bring to a boil. Then put on noodles.

# VEGETABLE CHOW MEIN, CANTONESE STYLE

## INGREDIENTS:

Egg noodles
Vegetables

## AMOUNT:

½ pound
Same as recipe for CHOW
 MEIN IN AMERICAN
 STYLE

## METHOD:

(1) If dried egg noodles are used, first put in boiling water for 3 minutes until noodles are soft. Then remove with a sieve and rince under running cold water for 1 minute. Let drain and dry for 1 hour.

(2) Use a bowl to prepare sauce with tapioca starch, ¾ cup water, 1 tablespoonful light soy sauce, and 2 table-spoonful oyster sauce.

(3) When ready to cook, heat up the wok to high heat. Add 3 tablespoonful peanut oil. When hot, put noodles to fry for 10 minutes. Keep turning until done. Remove to a plate and keep warm in an oven.

(4) Use 3 tablespoonful of oil in a hot wok, brown ginger, onion for a minute. Add all the vegetables, salt, and ¼ cup water. Stir mix and cover with a lid. Cook with high heat for 3 minutes. When done, pour in the sauce and bring it to a boil. Stir mix and put the mixture on top of egg noodles. Serve hot with chopped green onions if desired.

# VEGETABLE EGG ROLLS

## INGREDIENTS:

| INGREDIENTS: | AMOUNT: |
|---|---|
| Bean Sprouts | 1 pound |
| Cabbage | 1 cup, shredded |
| Bamboo Shoots | 1 cup, shredded |
| Onion | $\frac{1}{2}$ cup, shredded |
| Gourmet Powder | 1 teaspoonful |
| Light Soy Sauce | 2 tablespoonful |
| Celery | 1 cup, shredded |
| Cooking Wine | 1 tablespoonful |
| Tapioca Starch | 2 tablespoonful |
| Sesame Seed Oil | Drops |
| Egg Roll Wrappers | 1 package of 40 pieces |
| Ginger | 4 slices, then minced |

## METHOD:

(1) Use high heat, 2 tablespoonful of peanut oil, brown ginger and onions for a minute, then stir fry all the vegetables, salt, and gourmet powder for a minute. Cover with a lid. Cook with high heat for 5 minutes.

(2) When done, add starch and $\frac{1}{4}$ cup water into the mixture. Cook for another minute and remove to a tray for cooling.

(3) Fill a cup half with cold water and get a large plate for the wrapping.

(4) Take out wrapper one piece at a time and place it on top of the plate.

(5) Spoon 2 tablespoonful of filling into the centre of the wrapper, spread out lengthwise towards two corners. Pick up the lower corner and start folding it like a parcel. Smear the edges with water to make sure they are sealed. Make it long but narrow in a rectangular shape.

(6) Heat up 4 cups of oil in wok and deep fry egg rolls until golden brown. Serve hot with plum sauce.

# VEGETABLES, PICKLING

## INGREDIENTS: AMOUNT:

| INGREDIENTS: | AMOUNT: |
| --- | --- |
| Carrots | 2, skinned and cut into 2" sticks |
| Celery | 2 stalks, cut into 2" sticks |
| Cucumber | 1, cut into halves, then sliced diagonally at $\frac{1}{4}$" thick |
| Sugar | 1 cup |
| Vinegar, white | 6 - 8 cups |
| Sesame oil | Drops |

## METHOD:

(1)   Prepare about 8 cups of boiling water in a kettle.

(2)   Put all the vegetables in a large bowl. Pour boiling water over the vegetables until completely covered.

(3)   Immediately drain off the water.

(4)   Put in a cup of sugar, mix thoroughly with the vegetables.

(5)   Add vinegar to cover the vegetables and let soak for 20 minutes.

(6)   When done, drain off vinegar. Put in drops of sesame oil if desired, and put the vegetables to chill for 30 minutes or longer. The vinegar can be re-used for pickling or making sweet and sour sauce later.

# WON TON

## INGREDIENTS:

| INGREDIENTS: | AMOUNT: |
|---|---|
| Cabbage, shredded | 1 pound |
| Celery, shredded | 1 cup |
| Water Chestnuts | 1/2 cup, chopped fine |
| Gourmet Powder | 1 teaspoonful |
| Light Soy Sauce | 1 tablespoonful |
| Pepper | Dash |
| Egg | 1 large, beaten |
| Salt | 1/2 teaspoonful |
| Sesame Seed Oil | Drops |
| Won Ton wrappers | 1 package, thawed. Wrappers come in frozen state in a package of 50 pieces or more. |

## METHOD:

(1) Cook all vegetables in a wok. Then allow to cool.

(2) Properly thaw out wrappers but leave the package paper on. Mix all the ingredients together in a bowl.

(3) Fill a cup half with cold water. Use a large plate or paper towel as a tray for wrapping.

(4) Take out wrapper one at a time and place in the centre of the plate for wrapping.

(5) Pick up 1 teaspoonful of the mixture and place in the centre of the wrapper.

(6) Dip finger in water and smear lightly on the 4 edges of the wrapper.

(7) Pick up one corner and place it directly on top of the opposite corner. Press hard on edges and form a triangle.

(8) Smear lightly with water on the sloping edges of the triangle. Pick up the left lower corner and place directly on top of the upper corner. Press on finished edge.

(9) Repeat same procedure on the right lower corner. A pentagon shaped Won Ton is formed.

(10) Repeat the complete procedure until the mixture and wrappers run out.

# WON TON, DEEP FRIED

*Won Ton can be deep fried and served by itself or with plum sauce or sweet and sour sauce. They can be frozen and kept for a long time. For use, just heat up in oven. Or, for better result, deep fry once more in hot oil to maintain crisp condition.*

## METHOD:

In a wok heat up 6 cups of peanut oil. When oil is hot, put in Won Ton and deep fry until golden brown in colour. Remove from oil and place on top of paper towel to drain off excessive oil. Serve hot by itself or with plum sauce or sweet and sour sauce.

# WON TON IN SWEET AND SOUR SAUCE

## INGREDIENTS:            AMOUNT:

| INGREDIENTS | AMOUNT |
|---|---|
| Tapioca starch or cornstarch | 3 teaspoonful |
| Water | ¾ cup |
| Vinegar | ⅓ cup |
| Sugar | ¾ cup |
| Tomato paste | ½ of a small can |

## METHOD:

Mix all the ingredients in a saucepan. Stir constantly over high heat. Do not leave sauce unattended until the sauce is cooked. Pour over cooked Won Ton or arrange sauce in a serving bowl and invite your guests to dip Won Ton in during serving.

# ZUCCHINI IN BLACK BEAN SAUCE

## INGREDIENTS:

Zucchini

Salted Black Beans
Sugar
Tapioca Starch
Light Soy Sauce
Ginger roots, minced
Garlic, minced
Celery, diced
Onion, shredded
Sesame Seed Oil
Salt

## AMOUNT:

1-2 pounds, sliced diagonally
  into ¼ inch
3 tablespoonful, rinsed
1 teaspoonful
1 tablespoonful
1 talbespoonful
1 tablespoonful
1 tablespoonful
1 cup
1 cup
Drops
½ teaspoonful

## METHOD:

(1) Use a hot wok, add 2 tablespoonful of peanut oil, brown garlic, onion, ginger and black beans for 1 minute.

(2) Put in all vegetables, salt, 2 tablespoonful water, mix and cover with a lid. Cook with high heat for 3-5 minutes.

(3) When done, add starch solution (prepared with starch, ¼ cup water, soy sauce and sesame seed oil). Bring to a boil. Mix and serve hot.